THE AMERICAN LANDSCAPE

# GRAND CANYON

DWAYNE HICKS



PowerKiDS press.

Published in 2026 by The Rosen Publishing Group, Inc.
2544 Clinton Street, Buffalo, NY 14224



First Edition

Editor: Greg Roza
Book Design: Michael Flynn

Photo Credits: Cover, p. 1 Danita Delimont/Shutterstock.com; (series background) Omeris/Shutterstock.com; p. 5 Amanda Mohler/Shutterstock.com; p. 7 Radzas2008/Shutterstock.com; p. 9 Wisanu Boonrawd/Shutterstock.com; p. 11 (main) Kazuki Yamakawa/Shutterstock.com; p. 11 (inset) https://commons.wikimedia.org/wiki/File:4420_Grand_Canyon_Fossil_Bryozoan_in_Redwall_Limestone_(4749598980).jpg; p. 13 (main) Sergey Malomuzh/Shutterstock.com; p. 13 (inset) David G Hayes/Shutterstock.com; p. 15 (main) NPF Photography/Shutterstock.com; p. 15 (inset) Ronald Wilfred Jansen/Shutterstock.com; p. 17 RuslanKphoto/Shutterstock.com; p. 19 Nick Fox/Shutterstock.com; p. 21 dkroy/Shutterstock.com.

Library of Congress Cataloging-in-Publication Data

Names: Hicks, Dwayne, author.
Title: Grand Canyon / Dwayne Hicks.
Description: [Buffalo, New York] : PowerKids Press, [2025] | Series: The American landscape | Includes bibliographical references and index.
Identifiers: LCCN 2024035060 (print) | LCCN 2024035061 (ebook) | ISBN 9781499450866 (library binding) | ISBN 9781499450859 (paperback) | ISBN 9781499450873 (ebook)
Subjects: LCSH: Natural history–Arizona–Grand Canyon–Juvenile literature. | Geology–Arizona–Grand Canyon–Juvenile literature. | Grand Canyon (Ariz.)–Juvenile literature. | Grand Canyon (Ariz.)–Environmental conditions–Juvenile literature.
Classification: LCC QH105.A65 H53 2025 (print) | LCC QH105.A65 (ebook) | DDC 557.91/32–dc23/eng/20240731
LC record available at https://lccn.loc.gov/2024035060
LC ebook record available at https://lccn.loc.gov/2024035061

Manufactured in China



CPSIA Compliance Information: Batch #QSPK26. For Further Information contact Rosen Publishing at 1-800-237-9932.

# CONTENTS

## The Grandest of Canyons

A canyon is a deep, narrow valley, usually with a river flowing through it. The Grand Canyon is one of the largest canyons in the world! It's over 270 miles (435 km) long. The widest parts are 18 miles (29 km) wide. The Grand Canyon is over 1 mile (1.6 km) deep in some places!

## Where in the World?

The Grand Canyon is in Arizona. The Colorado River flows through it. This area, or piece of land, is dry and hot most of the year. There is a lake on both ends of the Grand Canyon. People made both Lake Powell and Lake Mead by building dams. These lakes are called reservoirs.

UTAH
Lake
Powell
NEVADA
Grand Canyon
Colorado River
Las Vegas
Lake
Mead
Colorado River
Flagstaff
ARIZONA
CALIFORNIA

## Erosion

The Grand Canyon formed over millions of years. It was shaped by **erosion**. The Colorado River carries away dirt and rocks as it flows through the canyon. Wind also helps shape the land of the canyon. After many years, erosion made the canyon deeper.

## Reading the Layers

Visitors to the Grand Canyon can see the many **layers** of rock that built up over time. Erosion cut through these layers and uncovered them. Scientists learn about Earth's past and how it formed by studying the layers. They have also found many **fossils** to study.

Grand Canyon fossil

## Canyon Plants

The top of the Grand Canyon is a mix of forests and prairie, or grassland. Natural springs feed hanging gardens along the canyon walls, especially in spring. At the bottom, where it's more like a desert, the river feeds plants such as cactuses, ferns, and mosses.

prickly pear cactus

ponderosa pines

# Canyon Animals

Mule deer are common near the rim of the Grand Canyon. Visitors may also see desert bighorn sheep, coyotes, and bison. Countless bugs, birds, and fish live there. Small **mammals** such as beavers, chipmunks, squirrels, rabbits, and bats live there too. There are also small **reptiles** such as snakes, lizards, frogs, and toads.

squirrel

collared lizard

## Native Americans

Native Americans have lived near the Grand Canyon for nearly 12,000 years. We know this because of very old **spear** points found there. The Hopi, Navajo, and Zumi are among the Native groups that still live here. Today, visitors can learn about Native American life through shows, **museums**, and more.

## Grand Canyon National Park

Europeans first came to the Grand Canyon in the mid-1500s. They mapped the land in 1858, the same year it was first studied by a scientist. **Tourism** began to grow in the early 1900s due to new railroads. In 1919, the Grand Canyon became a national park.

GRAND CANYON
NATIONAL PARK

## Visiting the Grand Canyon

Each year, more than 4 million people from around the world travel to the Grand Canyon! People can ride boats on the Colorado River, or ride mules down into the canyon. If you're brave enough, you can view the canyon from the Skywalk at Eagle Point!

# GLOSSARY

**erosion:** The wearing away of Earth's surface by wind, water, or ice.

**fossil:** A print or the remains of a plant or animal from long ago preserved in rock.

**layer:** One thickness of something lying over or under another.

**mammal:** Any warm-blooded animal whose babies drink milk and whose body is covered with hair or fur.

**museum:** A building where things, such as art and historical objects, are gathered and shown to the public.

**reptile:** A cold-blooded animal with thin, dry pieces of skin called scales.

**spear:** A long object made of a straight stick and a sharpened rock at the end used for hunting or war.

**tourism:** The business of drawing in tourists, or people traveling to visit another place.

# FOR MORE INFORMATION

## BOOKS

Leaf, Christina. *Grand Canyon National Park*. Minneapolis, MN: Bellwether Media, 2023.

O'Neal, Claire, and Tammy Gagne. *The Grand Canyon*. Mount Joy, PA: Curious Fox Books, 2024.

Walton, Kathryn. *Visit Grand Canyon National Park!* Buffalo, NY: Enslow Publishing, 2025.

## WEBSITES

**Easy Science for Kids: Grand Canyon—One of the Seven Wonders of the World**
*easyscienceforkids.com/all-about-the-grand-canyon*
Learn more interesting facts about the Grand Canyon.

**Kiddle: Grand Canyon Facts for Kids**
*kids.kiddle.co/Grand_Canyon*
Visit this website to read more about the Grand Canyon.

# INDEX